INTRODUCTION

WHAT IS A VOLCANO?

A volcano is an opening in the Earth's crust through which ash, magma (molten rock), rock fragments and gases escape from below the surface.

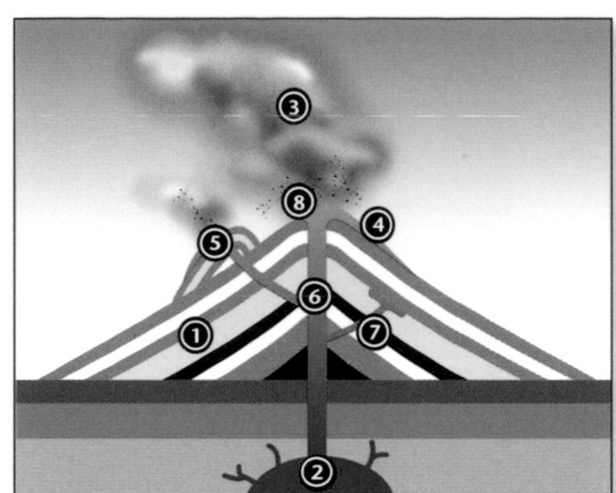

ANATOMY OF A COMPOSITE VOLCANO

1. **Cone** – Composed of layers of ash and lava that have hardened into rock.
2. **Magma Chamber** – Chamber beneath the volcano is filled with an explosive mixture of magma and gases.
3. **Ash Cloud** – Thick cloud of flour-like dirt and dust.
4. **Lava** – When liquified magma from inside the Earth reaches the surface, it becomes lava. Lava reaches the surface as a liquid flow or is ejected and cools in the air forming hardened rocks such as lapilli (little stones), blocks (angular chunks ranging from tennis ball-sized to house-sized pieces) or bombs (blobs of lava larger than 2.5 in./6 cm; if very viscous, they splatter on the ground like cow pies). Tephra is the term now used by volcanologists to refer to fragments of volcanic material of any size that is explosively ejected from a volcano.
5. **Lava Dome** – Roughly circular mound built by the slow extrusion of lava – like squeezing toothpaste out of a tube.
6. **Vent** – Chimney through which magma is ejected. Smaller side vents often branch off the main vent that may be circular openings, elongate fissures or small cracks.
7. **Dike** – Tubes or sheets of magma that cut through layers of rock and then cool and harden.
8. **Caldera** – Generally, a large, usually circular depression at the summit of a volcano which is formed when magma is withdrawn or rejected. The caldera at Kilauea – the world's most active volcano – fills and empties with lava on a regular basis.

Waterford Press produces reference guides that introduce novices to nature, science, travel and languages. Product information is featured on the website: **www.waterfordpress.com**

Made in the USA

To order, call 800-434-2555. For permissions, or to share comments, e-mail editor@waterfordpress.com.
For information on custom-published products, call 800-434-2555 or e-mail info@waterfordpress.com.

ISBN 978-1-58355-847-8 $7.95 U.S.

Volcanoes

SECOND EDITION

A Folding Pocket Guide to Volcanoes, Earthquakes, Hot Springs, Geysers & More

VOLCANOES Second Edition Kavanagh/Leung

TYPES OF VOLCANOES

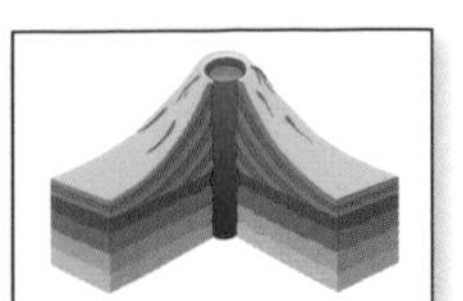

Cinder Cone – Small volcanoes with a single opening, bowl-shaped crater and steep sides formed of layers of lava and ash. Can evolve into a shield volcano over time (Paricutin, Cerro Negro, Mt. Zion, Lava Butte, Craters of the Moon).

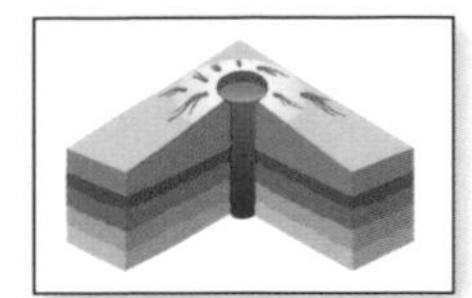
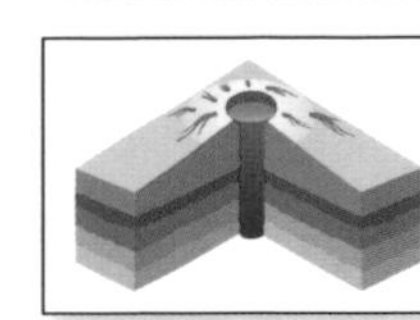

Shield Volcano – Large volcanoes with broad, gentle slopes formed by multiple lava flows. Magma has low viscosity and is prone to flowing, not exploding. Typical of oceanic volcanoes like those found in Hawaii.

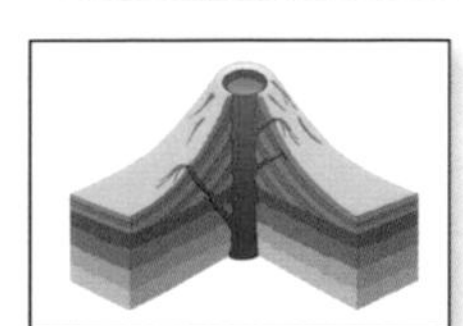
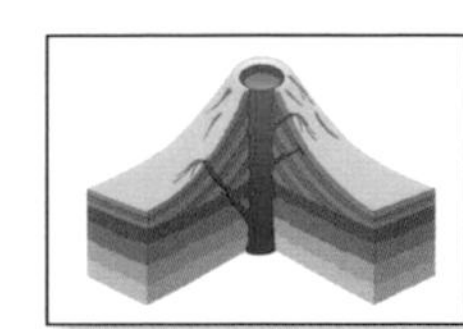

Composite or Stratovolcano – Tall, steep-sided mountains formed from alternating layers of lava, ash and cinders that are often found along convergent boundaries. Can be highly explosive (Mt. Vesuvius, Mt. Fuji, Mt. Rainier, Mt. Hood, Mt. Baker).

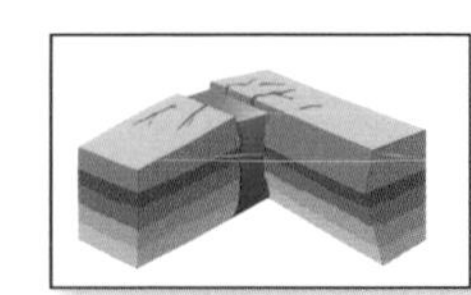

Rift/Fissure Volcano – A long crack in the Earth's surface through which magma erupts and lava spreads far and wide. Rifts are bigger than fissures. Typically a result of divergent plates on mid-ocean ridges (Iceland, Africa).

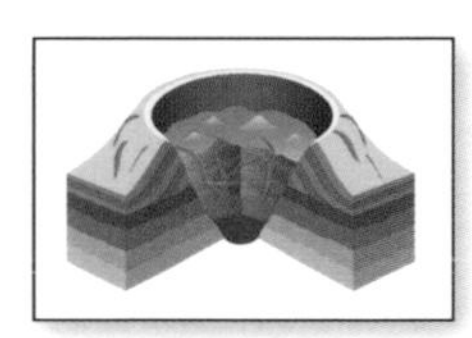
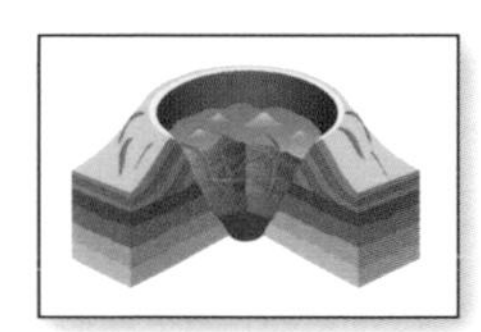

Supervolcano Caldera – Large oval-shaped depression formed when the ground fell in after the eruption of magma from large underground magma chambers (to 30 miles/50 km wide). Caldera-forming eruptions, like the one that formed Yellowstone, are the largest eruptions on Earth.

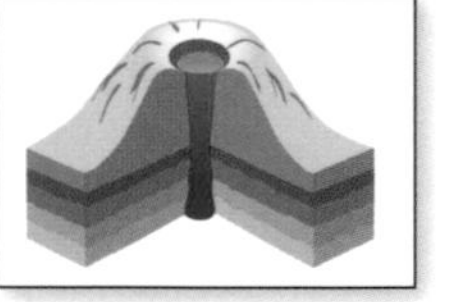
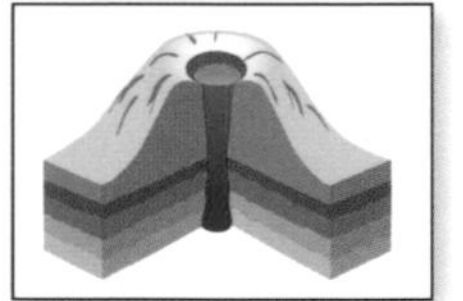

Dome Volcano – Steep-sloped, mound-shaped protrusion (a mountain or a volcano feature) created by multiple eruptions of highly viscous lava. Can be highly explosive (Mammoth Mountain, Novarupta, Mt. Lassen).

VOLCANIC HAZARDS

- **Lava Flows**
- **Airborne Projectiles** (tephra).
- **Volcanic Clouds** of ash and poisonous gases.
- **Pyroclastic Flows** – Ground-hugging avalanches of hot ash, debris and gases.
- **Lahars** – Volcanic mudslides of rocks, mud and water.
- **Debris Avalanches** – Rock or snow slides.
- **Acid Rain**

Lava flows obliterate (almost) everything in their path.

VOLCANIC ROCKS

Molten lava erupts at the surface of the Earth at temperatures between 700°C - 1,200°C (1,300°F - 2,200°F). Lavas differ in form and texture. When lava flows stop moving, they form igneous rocks. Depending on the lava's viscosity (stickiness), it hardens in different ways. There are three types of lava flows: pillow lava (forms puffy lava rocks underwater), aa and pahoehoe.

Aa Flow
Blocky, jagged lava is produced by slow, viscous flows.

Pahoehoe Flow
Ropey, wrinkly lava is produced by more fluid flows.

Igneous rocks are formed when lava is either ejected directly onto the earth's surface (extrusive rock) or pushes up near the surface and cools beneath it (intrusive rock). Intrusive rocks tend to crystallize more slowly and have larger minerals (crystals) than extrusive rocks. The composition of the magma dictates the types of minerals that can form.

EXTRUSIVE IGNEOUS ROCKS

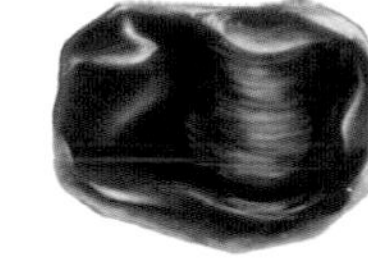
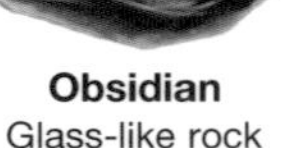

Obsidian
Glass-like rock

Basalt
Hard, black rock

Pumice
Light-colored rock riddled with holes. Floats.

INTRUSIVE IGNEOUS ROCKS

Granite
Typically white to gray rock has a 'salt and pepper' mix of large crystals.

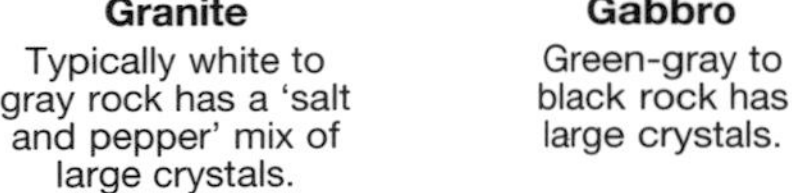

Gabbro
Green-gray to black rock has large crystals.

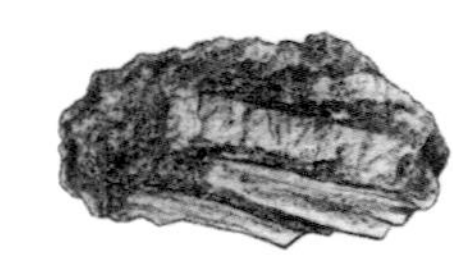

Granite Porphyry
Similar to granite but with much larger crystals.

PYROCLASTIC FLOW

This terrifying volcanic hazard occurs when an eruption ejects a cloud of debris so dense with fragments that it quickly falls back to Earth as a ground-hugging avalanche of hot ash, debris and poisonous gases. With temperatures as hot as 1,500°F (815°C), the cloud travels at speeds of up to 150 mph (240 kph) and burns and flattens everything in its path. When Mt. Vesuvius erupted in 79 AD, the residents of Pompeii were killed by a pyroclastic surge before the town was buried in ash and lava.

WHERE VOLCANOES ARE FOUND

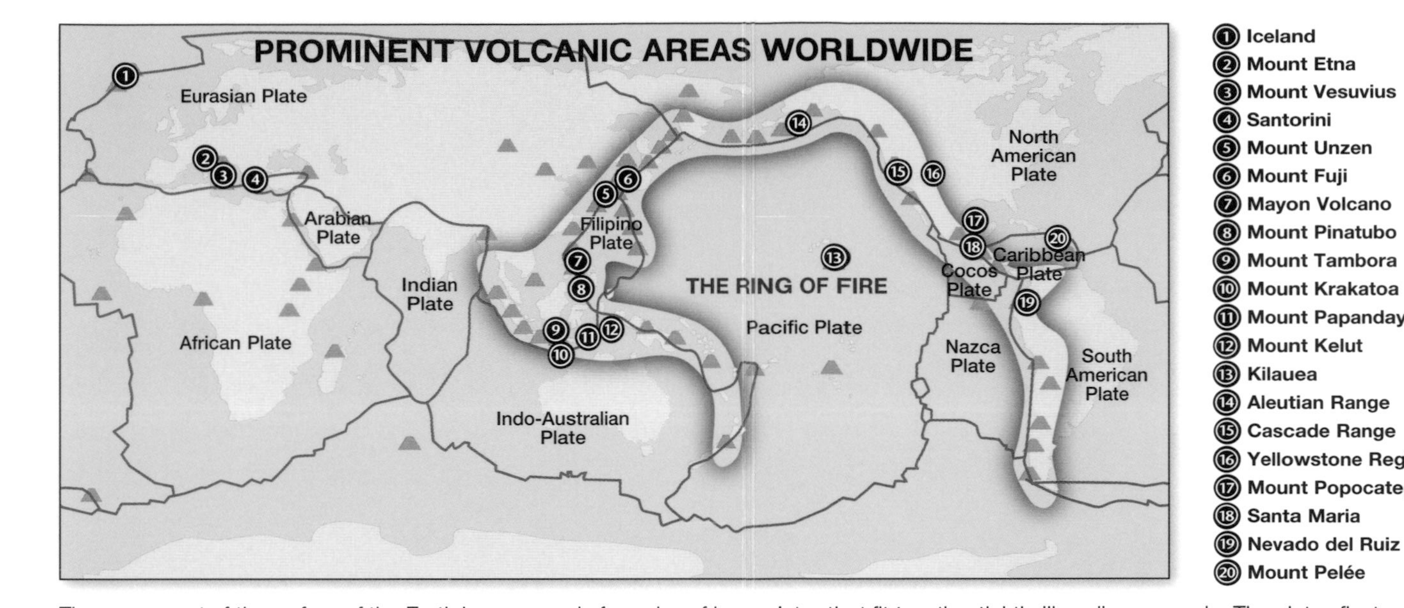

1. Iceland
2. Mount Etna
3. Mount Vesuvius
4. Santorini
5. Mount Unzen
6. Mount Fuji
7. Mayon Volcano
8. Mount Pinatubo
9. Mount Tambora
10. Mount Krakatoa
11. Mount Papandayan
12. Mount Kelut
13. Kilauea
14. Aleutian Range
15. Cascade Range
16. Yellowstone Region
17. Mount Popocatepetl
18. Santa Maria
19. Nevado del Ruiz
20. Mount Pelée

The upper crust of the surface of the Earth is composed of a series of huge plates that fit together tightly like a jigsaw puzzle. The plates float around on a fluid-like solid (asthenosphere) and are constantly pressing together; pulling apart; and sliding over, under or grinding by each other. The boundaries of plates are where mountain ranges are formed and the majority of earthquakes and volcanic eruptions occur. Some of the most active volcanoes in the world occur along plate edges in the Pacific region, an area known as the 'Ring of Fire'.

VOLCANO/EARTHQUAKE/TSUNAMI

Volcano – A vent through which magma, ash, or volcanic gases erupt (see types of volcanoes).

Earthquake – The sudden and violent shaking of the ground resulting from rocks fracturing within the Earth's crust. When rocks fracture, shock waves travel away from the source of the fracture (epicenter). The shock waves are usually most severe near the source of the fracture. Earthquakes can be tectonic, caused by the movement of tectonic plates, or volcanic, caused by the movement of volcanic fluids.

Japan Earthquake 2011

Tsunami – A series of huge, fast-moving waves that are a result of a large, abrupt displacement of the sea floor, usually caused by earthquakes or volcanic eruptions. Waves can be up to 280 ft. (80 m) high and travel at speeds of up to 370 mph (600 kph). On average, about two tsunamis a year occur that are large enough to inflict damage; once every 10-15 years a destructive, ocean-wide tsunami occurs.

VOLCANO/EARTHQUAKE/TSUNAMI CONNECTION

In February of 2011, the Kilauea caldera in Hawaii was full of bubbling lava. In early March, an earthquake near Japan caused a huge tsunami in that region that, at the same time, caused the Kilauea caldera to drain of lava.

TOP TEN DEADLIEST ERUPTIONS

1. **Mount Tambora,** Indonesia – 1815 – Death Toll: 60,000 est. Series of eruptions caused global climactic effects and widespread famine.
2. **Mount Krakatoa,** Indonesia – 1883 – Death Toll: 36,000 Series of gigantic explosions were so powerful that they burst the eardrums of sailors 25 mi. (40 km) away. Most of the lives lost were due to the resulting tsunamis.
3. **Mount Vesuvius,** Italy – 79 AD – Death Toll: 33,000 Ash and lava buried the cities of Pompeii and Herculaneum.
4. **Mount Pelée,** West Indies – 1902 – Death Toll: 29,000 Levelled and burned the city of St. Pierre. Only two people survived.
5. **Nevado del Ruiz,** Colombia – 1985 – Death Toll: 23,000 Massive mudslide buried the city of Armero.
6. **Mount Unzen,** Japan – 1792 – Death Toll: 15,000 Several lava domes collapsed, creating a deadly tsunami.
7. **Laki Volcanic System,** Iceland – 1783 – Death Toll: 9,350 Clouds of poisonous gas killed 50% of the nation's livestock and 25% of the people.
8. **Santa Maria,** Guatemala – 1902 – Death Toll: 6,000 Explosion ejected debris over 100,000 sq. mi. (260,000 sq. km).
9. **Mount Kelut,** Indonesia – 1919 – Death Toll: 5,000 Eruption created massive mudslide.
10. **Mount Papandayan,** Indonesia – 1772 – Death Toll: 3,000 Debris avalanche destroyed 40 villages.

PLATE TECTONICS

Over the last 250 million years, the Earth is believed to have evolved from a single land mass into its current form. Over millions of years, the supercontinent Pangaea broke into a series of floating plates and these drifted apart to form the world as we know it today. Where the plates floated into each other, mountain ranges were formed. Proof of the theory of continental drift is the presence of identical fossil specimens and rock formations on the coastlines of continents that are now great distances apart.

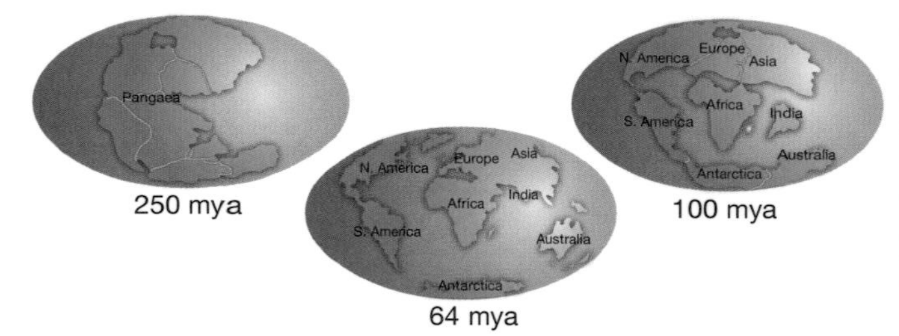

250 mya

100 mya

64 mya

HOW ARE VOLCANOES FORMED?

Volcanoes are formed when plates pull apart from each other, slide over/under each other, or when a plate travels over a hot spot.

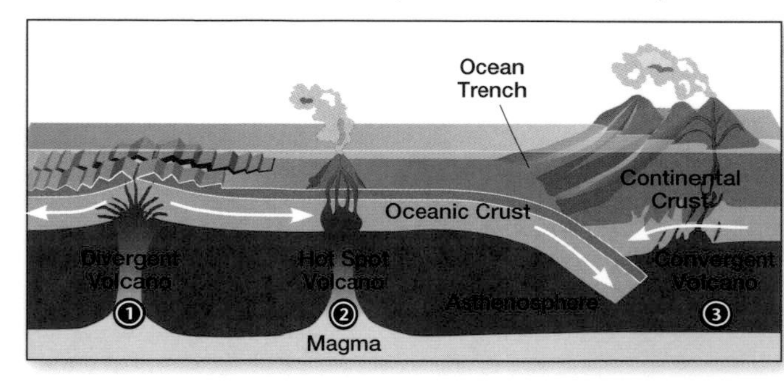

Volcanoes are formed in three ways:

1. **Along Divergent (constructive) Boundaries** – At mid-ocean ridges, two plates spread apart from each other, creating fissures and vents in the crust that magma rises through.
2. **Over Hot Spots** – As a plate moves over a hot spot, magma erupts through the plate. These mantle plumes are relatively fixed in space, and magma continues to erupt through the crust as the plate moves over it. This is the hypothesis for how volcanic island chains are formed. For example, the Hawaiian Islands have the oldest, extinct volcanic islands at the west end, and the youngest, active volcanic islands at the east end. Further proof of this theory is that a new active undersea volcano – Lo'ihi Seamount – is erupting off the southeast coast of the volcanically active Big Island of Hawaii, and is expected to surface in 10,000-100,000 years.
3. **Along Convergent (destructive) Boundaries** – The leading edge of one plate slides under another. As a result of friction, pressure and plate material remelting, earthquakes and volcanoes are common near these boundaries. They typically occur between an oceanic plate and a continental plate.

PLATE TECTONICS

TRANSFORM FAULTS

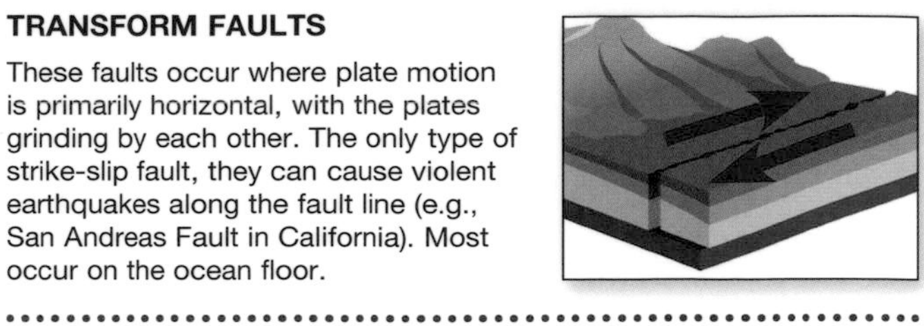

These faults occur where plate motion is primarily horizontal, with the plates grinding by each other. The only type of strike-slip fault, they can cause violent earthquakes along the fault line (e.g., San Andreas Fault in California). Most occur on the ocean floor.

CONTINENTAL DRIFT

Continents are continually on the move, even today. North America, for example, moves west at about one inch (3 cm) per year. In about 63,000 years (a blink in geologic time) it will be a mile (1.6 km) further west.

OTHER VOLCANIC FEATURES

Geysers – A spring that intermittently discharges water and vapor. Generally, surface water works its way down to where it contacts hot rocks. The boiling water builds up pressure, causing water and steam to be ejected through a surface vent. The famous Yellowstone geyser – Old Faithful – erupts approximately every 78 minutes.

Old Faithful

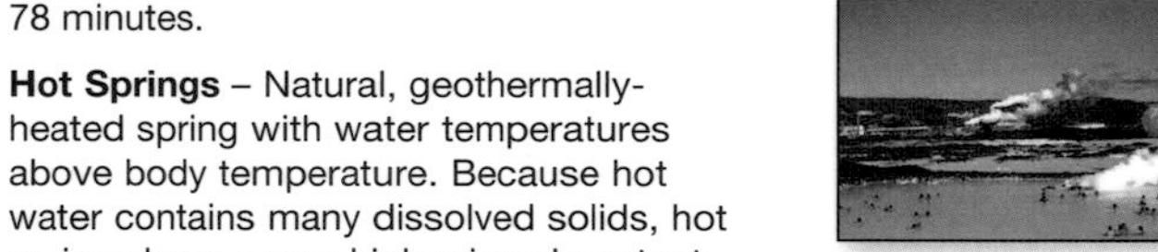

Hot Springs – Natural, geothermally-heated spring with water temperatures above body temperature. Because hot water contains many dissolved solids, hot springs have a very high mineral content and are believed to be therapeutic for a variety of ailments.

Blue Lagoon, Iceland

Mud Pot – Also known as paint pots, these are depressions filled with mud that is usually boiling. They occur in geothermal areas that have a shortage of water. The edges of the pot may be stained reddish by iron compounds in the mud.

Boiling mudpot

Fumarole – A volcanic vent that emits fumes of steam and gases. They occur when hot rocks or magma come in contact with groundwater. They may appear as small vents or as long cracks and may persist for decades or centuries.

Steaming fumaroles

VOLCANO TERMS

Active – A volcano currently erupting or is likely to do so in the near future.

Dormant – A volcano that is resting and is expected to erupt again.

Extinct – A volcano that is not expected to erupt again.

HOW VOLCANOES ARE MONITORED

Scientists that study volcanoes are called volcanologists. While all eruptions are unique events, volcanologists can use a number of tools and methods to attempt to predict how and when a volcano may erupt.

Volcanologists at work

SIGNS OF AN IMPENDING ERUPTION

1. **Seismic Activity** – Movement of fluids beneath the Earth's crust can generate a series of tremors or small earthquakes. Volcano seismologists watch for these types of earthquakes because they could indicate the movement of magma. Ideally, a series of 6-8 seismographs are positioned around a volcano to allow scientists to determine the size, number and types of tremors.
2. **Surface Deformation** – A variety of surveying techniques and instruments are used to measure surface deformation including global positioning systems, tiltmeters and electronic distance meters.
3. **Gases** – The amount and type of gas emitted by a volcano increases as magma nears the surface. Water (steam) and gases such as carbon dioxide and sulphur dioxide are often released. Devices such as ultraviolet spectrometers can measure sulphur dioxide concentrations.
4. **Temperature** – Changes in the temperature of volcanic springs and vents can indicate impending eruptions.
5. **Satellites** – Allow large areas to be monitored and thermally scanned.
6. **Other Volcanoes** – The activity of other volcanoes are studied in an attempt to uncover trends and patterns.
7. **History** – Information on past eruptions can help scientists attempt to predict the type of eruption to expect.

Seismograph

SUCCESSFUL PREDICTION

The 2000 eruption of Mount Popocatepetl was accurately predicted and over 50,000 people were successfully evacuated prior to the eruption – the most violent in 1,200 years.

Mount Krakatoa

Indonesian volcano was produced by a submarine eruption in 1927. It grows at an average rate of 23 ft. (7 m) each year!

FACT

There are at least 1,500 active volcanoes in the world. For more information, see the Smithsonian Global Volcanism Program (volcano.si.edu) and the U.S. Geological Survey (volcanoes.usgs.gov).

PROMINENT VOLCANIC AREAS

USA

There are approximately 170 active volcanoes in the United States. Most of these volcanoes are located on the Ring of Fire.

Mount Redoubt – Alaskan stratovolcano is likely the most active North American volcano. Last major eruption occurred in 1990.

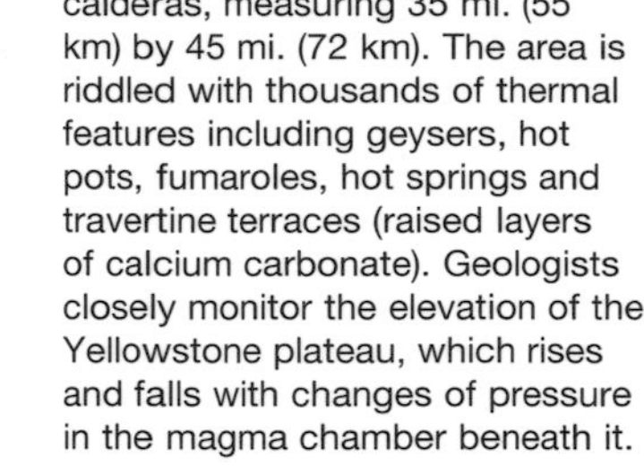

Yellowstone Volcano – Also known as the Yellowstone Super-volcano, it is one of the world's largest and most active calderas, measuring 35 mi. (55 km) by 45 mi. (72 km). The area is riddled with thousands of thermal features including geysers, hot pots, fumaroles, hot springs and travertine terraces (raised layers of calcium carbonate). Geologists closely monitor the elevation of the Yellowstone plateau, which rises and falls with changes of pressure in the magma chamber beneath it.

Grand Prismatic Spring, Yellowstone

Mount Baker – Large stratovolcano (10,780 ft./3,285 m) in northern Washington is the second-most active in the North Cascades after Mount St. Helens. Heavily glaciated, it receives up to 95 ft. (29 m) of snow each year. Last erupted in 1880.

Mount Rainier, Seattle

Mount Rainier – Dormant stratovolcano 54 mi. (87 km) from Seattle is the most heavily glaciated mountain in the North Cascade Range. Considered extremely dangerous due to the potential mudslides that could be triggered by an eruption.

Mount Hood, Portland

Mount Hood – Dormant stratovolcano 50 mi. (80 km) from Portland last erupted in 1866. Represents the same potential danger as Mount Rainier.

Mount St. Helens

Mount St. Helens – The most active volcano in the contiguous U.S. In 1980, the largest landslide in recorded history triggered a violent eruption that blew ash 15 mi. (24 km) into the air and ejected debris over 900 mi. (1,440 km) away.

Crater Lake

Crater Lake – Gigantic eruption created a crater 6 mi. (10 km) wide. The lake is 1,950 ft. (594 m) deep, the deepest in the United States. The caldera rim is up to 8,000 ft. (2,400 m) in elevation.

Sunset Crater

Sunset Crater – Symmetrical cinder cone is over 8,000 ft. (2,440 m) in elevation and 400 ft. (122 m) deep. Massive eruption around 1100 AD ejected tephra more than 800 sq. mi. (2,100 sq. km).

PROMINENT VOLCANIC AREAS

Lassen Peak – One of the largest plug dome volcanoes in the world. A series of spectacular eruptions occurred between 1914 and 1917. Today, the area surrounding the peak is still active with mud pots, fumaroles and hot springs.

Lassen Peak

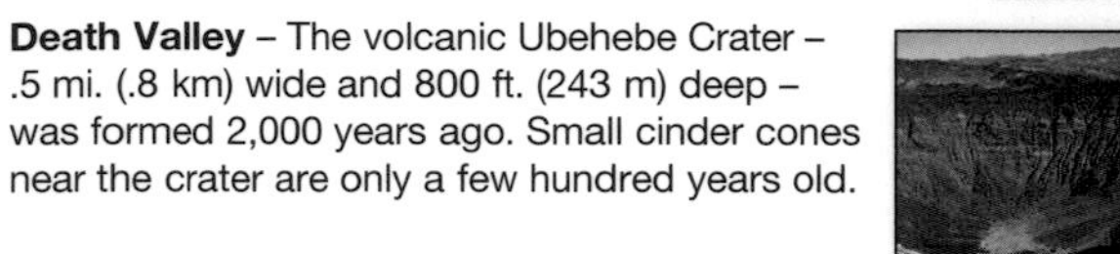

Death Valley – The volcanic Ubehebe Crater – .5 mi. (.8 km) wide and 800 ft. (243 m) deep – was formed 2,000 years ago. Small cinder cones near the crater are only a few hundred years old.

Ubehebe Crater

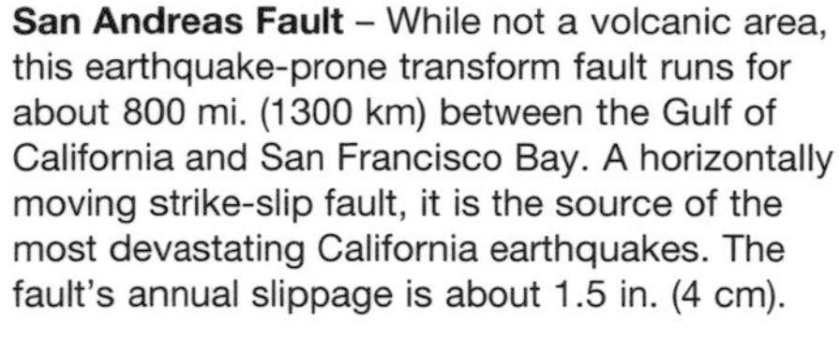

San Andreas Fault – While not a volcanic area, this earthquake-prone transform fault runs for about 800 mi. (1300 km) between the Gulf of California and San Francisco Bay. A horizontally moving strike-slip fault, it is the source of the most devastating California earthquakes. The fault's annual slippage is about 1.5 in. (4 cm).

San Andreas Fault

Mount Shasta – Stratovolcano is the second highest peak in the Cascade Range at 14,180 ft. (4,322 m). It consists of four overlapping volcanic cones.

Mount Shasta

SPOTLIGHT: THE BIG ISLAND OF HAWAII

Almost all of the active volcanoes in Hawaii are found on the Big Island of Hawaii, the youngest of the Hawaiian islands. East Maui Volcano, commonly known as Haleakala, on the island of Maui, is the only other Hawaiian volcano to have erupted since the late 1700s.

Kohala
Mauna Kea
Hualalai
Mauna Loa
Kilauea
The Big Island

Mauna Loa – This shield volcano makes up more than half of the island. Considered an active volcano, it has erupted 33 times since 1843 and as recently as 1984.

Mauna Kea – 13,796 ft. (4,205 m) – Hawaii's tallest mountain is the tallest mountain from top to bottom found on Earth, rising about 33,500 ft. (10,200 m) from the sea floor. Dormant, it last erupted 4,000 years ago.

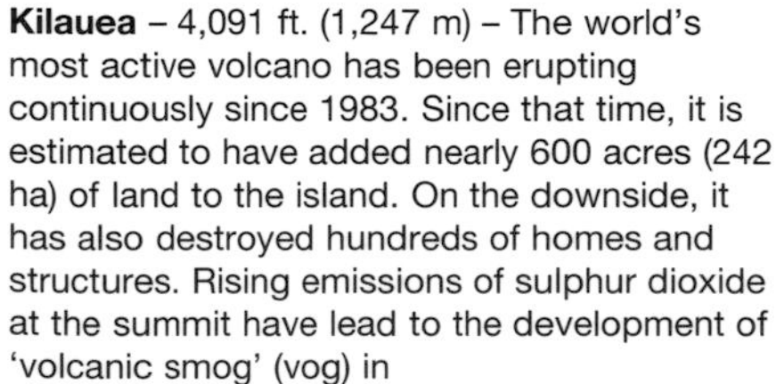

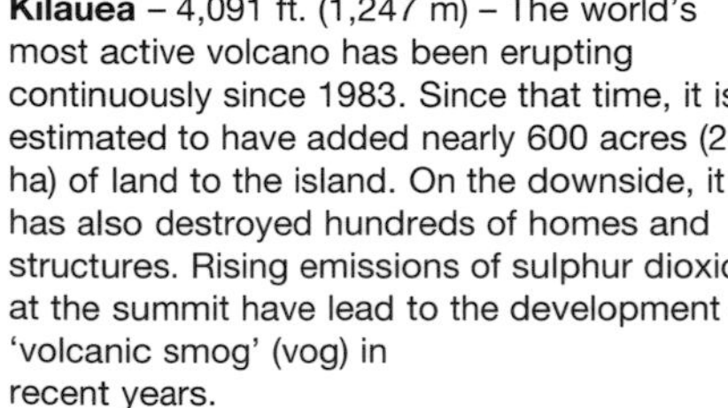

Kilauea – 4,091 ft. (1,247 m) – The world's most active volcano has been erupting continuously since 1983. Since that time, it is estimated to have added nearly 600 acres (242 ha) of land to the island. On the downside, it has also destroyed hundreds of homes and structures. Rising emissions of sulphur dioxide at the summit have lead to the development of 'volcanic smog' (vog) in recent years.

Kohala – 5,480 ft. (1,670 m) – The oldest volcano on the island is estimated to have last erupted 120,000 years ago. Extinct.

Hualalai – 8,271 ft. (2,521 m) – Dormant, it last erupted in 1801.

PROMINENT VOLCANIC AREAS

WORLDWIDE

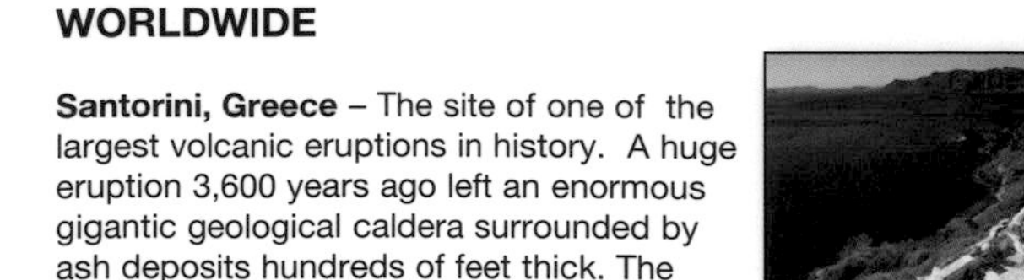

Santorini, Greece – The site of one of the largest volcanic eruptions in history. A huge eruption 3,600 years ago left an enormous gigantic geological caldera surrounded by ash deposits hundreds of feet thick. The eruption was estimated to have ejected debris 100 mi. (160 km) into the air.

Santorini Caldera

Mount Vesuvius, Italy – This stratovolcano is the only active volcano on the European mainland. An enormous 79 AD eruption buried the cities of Pompeii, Herculaneum and Stabiae under a wave of lava. Considered one of the most dangerous volcanoes in the world given that 3 million people live in its vicinity.

Mount Vesuvius, Naples

Mount Etna, Italy – One of the world's most active volcanoes, it has erupted 190 times in the last 3,500 years. Has three prominent craters at its summit. On the sides of the volcano are more than 300 vents that were created by flank eruptions.

Mount Etna

Eyjafjallajökull/Grímsvötn, Iceland – There are about 33 active volcanoes in Iceland. Large eruptions in 2010-2011 sent up dense clouds of ash that disrupted air traffic throughout Europe.

Mount Popocatepetl, Mexico – Active stratovolcano near Mexico City is the second highest peak in Mexico at 17,800 ft./5426 m. Major recent eruptions have occurred in 2004, 2003, 1994-1995, 1947 and 1942-1943. The largest of the eruptions were in 1996-2003.

Mount Popocatepetl

Mount Pinatubo, Philippines – Colossal 1991 eruption – the second-largest of the 20th century – ejected more than a cubic mile (5 cubic km) of volcanic material and lowered global temperatures by 1°F (0.5°C) for one year. Thousands of buildings were destroyed by pyroclastic flows and lahars.

Mount Pinatubo Crater

Mayon Volcano, Philippines – Perfectly conical stratovolcano is the most active volcano in Indonesia. Has erupted 49 times since 1616. Was created by convergent plates in the 'Pacific Ring of Fire'.

Mayon Volcano

Mount Fuji, Japan – Perfectly symmetrical, active stratovolcano is the largest mountain in Japan (3,776 ft./12,390 m). Located 80 miles (130 km) from Tokyo and its 13 million residents.

Mount Fuji